GRADE

MATH Trailblazers®

A BALANCED MATHEMATICS PROGRAM INTEGRATING SCIENCE AND LANGUAGE ARTS

Adventure Book

THIRD EDITION

KENDALL/HUNT PUBLISHING COMPANY
4050 Westmark Drive Dubuque, Iowa 52002

A TIMS® Curriculum
University of Illinois at Chicago

MATH TRAILBLAZERS®

Dedication

This book is dedicated to
the children and teachers who
let us see the magic in their classrooms
and to our families who wholeheartedly
supported us while we searched for
ways to make it happen.

The TIMS Project

UIC The University of Illinois at Chicago

The original edition was based on work supported by the National Science Foundation under grant No. MDR 9050226 and the University of Illinois at Chicago. Any opinions, findings, and conclusions or recommendations expressed in this publication are those of the authors and do not necessarily reflect the views of the granting agencies.

ISBN 978-0-7575-3490-4

Printed in the United States of America
1 2 3 4 5 6 7 8 9 10 11 10 09 08 07

Table of Contents

Look Around You

Look around you—
What do you see?
A room that's a brand-new
Place to be!

Today, we're going to play a game
To see how many numbers you can name.
We're going to count things, big and small.
And 1, 2, 3—it's fun for all!

Windows, windows in the wall,
How many are there all in all?
We can count them to find out.
And when we know, we'll give a shout!

Are there enough seats to go around?
How can the answer to this be found?

We can count them to find out.
And when we know, we'll give a shout!

How many pencils in my cup?
Take them out, and hold them up.

We can count them to find out.
And when we know, we'll give a shout!

How many steps to make your way
Across the room, right now, today?

We can count them to find out.
And when we know, we'll give a shout!

Look around you—
What do you see?
A room that's a brand-new
Place to be!

Now it's YOUR turn to look around
And count some objects YOU have found.

Look around you—
What do you see?
That counting's as easy
As 1, 2, 3!
And when you've counted and found out,
Give a great, big, well-earned shout!

I've Got a Little List

Hey, Joe! My group collected some interesting data from the *Colors* experiment.
Color
red 12
green 6
yellow 7
orange 2

May I see your data?
Okay, Lucy.

What's that?
My data.
There are no color names!
How do you know which is which?
I can remember!

Which is yellow?
That one— the 11!
9
11
2
1

And what goes on the bottom?
9
11
2
1

I don't remember...

I think you're going to have a problem with your homework.

I guess so.
I need to get home. Good luck!

Hi, honey, what do you have?
The data from my experiment.

Great! But why the tears?
Not so great. I didn't write down the color names. Now, I can't do my homework.

I've got an idea. Why don't you call your partner?
Okay, I sure hope she can help.

Hi, Jill. I really need your help. I forgot to label the data in my table. What do you have?

8 red instead of 9—oh, boy!
11 yellow,
2 green,
1 orange.
Thanks, Jill, for all your help!

Honey, why didn't you write in the colors at school?
I thought I would remember them when I got home.
Color
red
yellow
green
orange
11
2
1

I guess it wasn't such a good idea.
Many times, I've stood in the aisle at the grocery store, and I couldn't remember what I was supposed to buy.
EGGS
MILK
COOKIES
BREAD
JAM
BUTTER

So, I keep a list.
Then I bring my list to the store.
EGGS
MILK
COOKIES
BREAD
JAM
BUTTER

We all forget, dear.
You're telling me!

Betty Builds a Better Racer

Wow!
Neat, Betty!
Hi, Tom. My mom and I built my racer. It's for the soap box derby on Saturday.
5

I've got to test it to make it better.
How?

See the wheels and wood?
I'm going to try different wheels and streamline the body.

We don't need those wheels yet. Grab the big board, and I'll show you what to do.

Put the board here.
It's a ramp!

Let's line the ramp up with a sidewalk crack. Then we can put a straight line on the ramp.
I get it! The car will always start at the same spot.

We can use this chalk to number the sidewalk spaces.

We'll let it roll until it stops.
Right! Let it go!
5

See, it went almost 5 spaces. Write that down in a data table.
Just like in class.
5

What's going on?
Betty is building a better racer!
5

Do you think these wheels are better?
Make sure the back wheels are even with the line, Tom.
5

Great, Betty! It went 6 spaces.
I wrote that down. Now, I want to streamline the body.
6

See how boxy it is?
I want it sleek like that car.
5

With better wheels and a sleeker body, it should really go!
We can nail in boards here.
Here, too!
I get it! Now, the air will slide over the car.
5

Wow! 9 spaces!
I recorded the data. Now, the racer needs a good sanding!
9

The day of the race…
Last up is number 5, Betty! To win, she has to go farther than flag 6. She's ready.
Yes!
5
1
2

Go, Betty, go!
2
3
4
5
Can I do it?
5

7
8
9
We won! Almost 8 flags! A new record!
Betty did build a better racer!
5

The Midnight Visit

Hi, Emilio!
Hi, Mom.

What do you have?
My homework assignment.

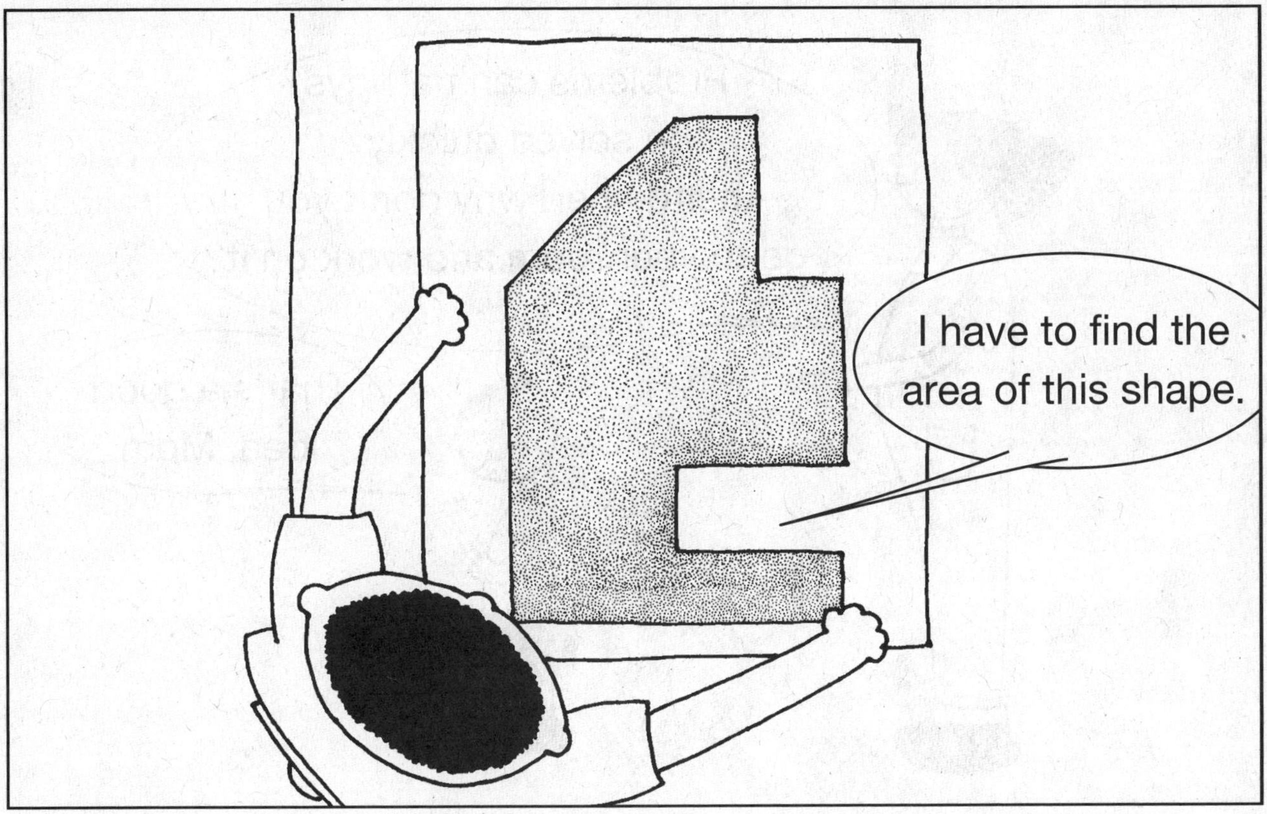
I have to find the area of this shape.

Do you know how?
We worked on area in school today, but I'm not sure exactly what I need to do.

Problems can't always be solved quickly. After dinner, why don't you go to your room and work on it?
That's a good idea, Mom.

Well, I've worked a long time on this problem, but I'm not finished yet.

I need to sleep on the problem.
I'll try to finish in the morning.
If I can't figure it out, I'll ask my teacher for help tomorrow.
Emilio was really working hard tonight to solve a problem.
What was he trying to do?

He has to find the area of this shape.
I know how. Follow me.

Why are we going to the refrigerator?
You'll see. Help me open it.

I'm throwing down the cheese slices. Get a knife!
CHEESE

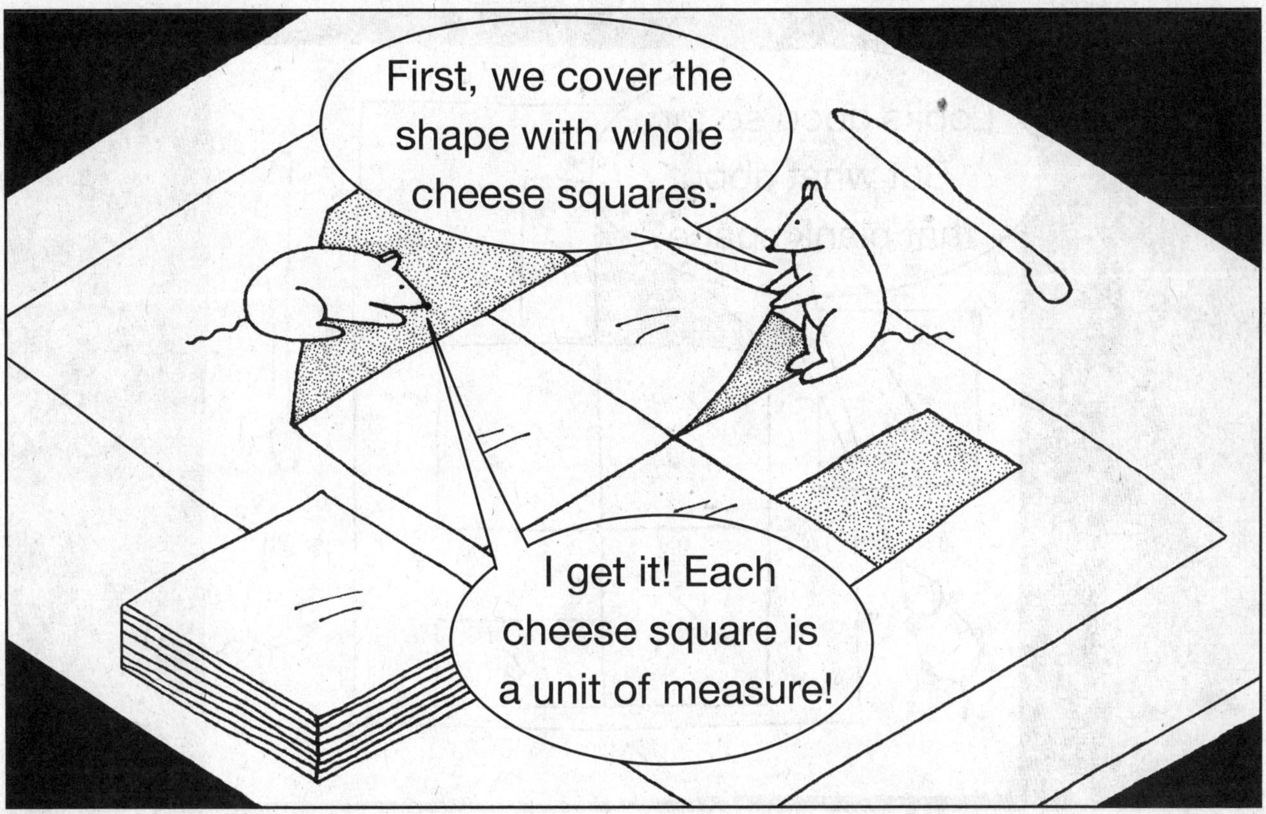
First, we cover the shape with whole cheese squares.
I get it! Each cheese square is a unit of measure!

Now, cut one in half so we have two pieces to fit in the smaller areas. Cut the slice in two to make two rectangles.
Okay.

Looks good so far. But what about that blank space?

Simple. We cut one on the diagonal.

That fits here.
Okay! Let's eat the other half.

We're done!
Fantastic! How many slices?

The next morning…
What is this?
Oh! I see! The cheese covers the area of the shape on my homework. Hey, I remember now! That's a way to find the area. You can count the pieces it takes to cover it. There are $4\frac{1}{2}$ square slices of cheese. I'm going to cover this and take it to school!
Sally, how much area did you find for last night's homework?
Square inches?
Oh, no!
Mine is 72 square inches.
Mine too!

Emilio, what area did you find?
Mine is $4\frac{1}{2}$ square cheese slices.
Yes, cheese. What's wrong with cheese?
Cheese?
Cheese? You could use cheese?

Well, we could use cheese, provided everyone used cheese.

Square inches are okay, and square cheese slices are okay. But not both. We all need to use the same unit.
Cheese smells. Square inches are better.

Well, since most of you used square inches, we'll make that our unit of measure. Good job, class!
We did a good job too, but I hope Emilio remembers how to do his homework tonight!

It's Sunny in Arizona

Chicago, Illinois, in January
Look at all that snow!
We never see the sun anymore.

I just got a letter. Your Aunt Carol has invited us to visit her in Tucson, Arizona.
Should we go?
BULLS

BULLS
YES!

Where is Arizona?
I don't know.

Arizona is in the southwest of our country.
We're going to drive.
BULLS

You kids can keep track of the trip on the map.
Yeah!
Close the window, please. It's cold.
KRT680

MAP OF THE UNITED STATES OF AMERICA
North
East
West
South
Westward, ho!

Which way do we head now?
Straight south.

It's starting to rain.

Yes, it gets warmer as we go farther south.
The snow turns to rain.
It's still cloudy and cold.

Springfield, Illinois
WYOMING
SOUTH DAKOTA
IOWA
NEBRASKA
Chicago
ILLINOIS
UTAH
COLORADO
MISSOURI
St. Louis
KANSAS
Flagstaff
Mississippi River
ARIZONA
Albuquerque
Amarillo
OKLAHOMA
Oklahoma City
ARKANSAS
NEW MEXICO
Tucson
TEXAS
MISSISSIPPI
LOUISIANA
He was the 16th president—a great man.
That's Abraham Lincoln's house.

Is that Abe up there?
Scary!

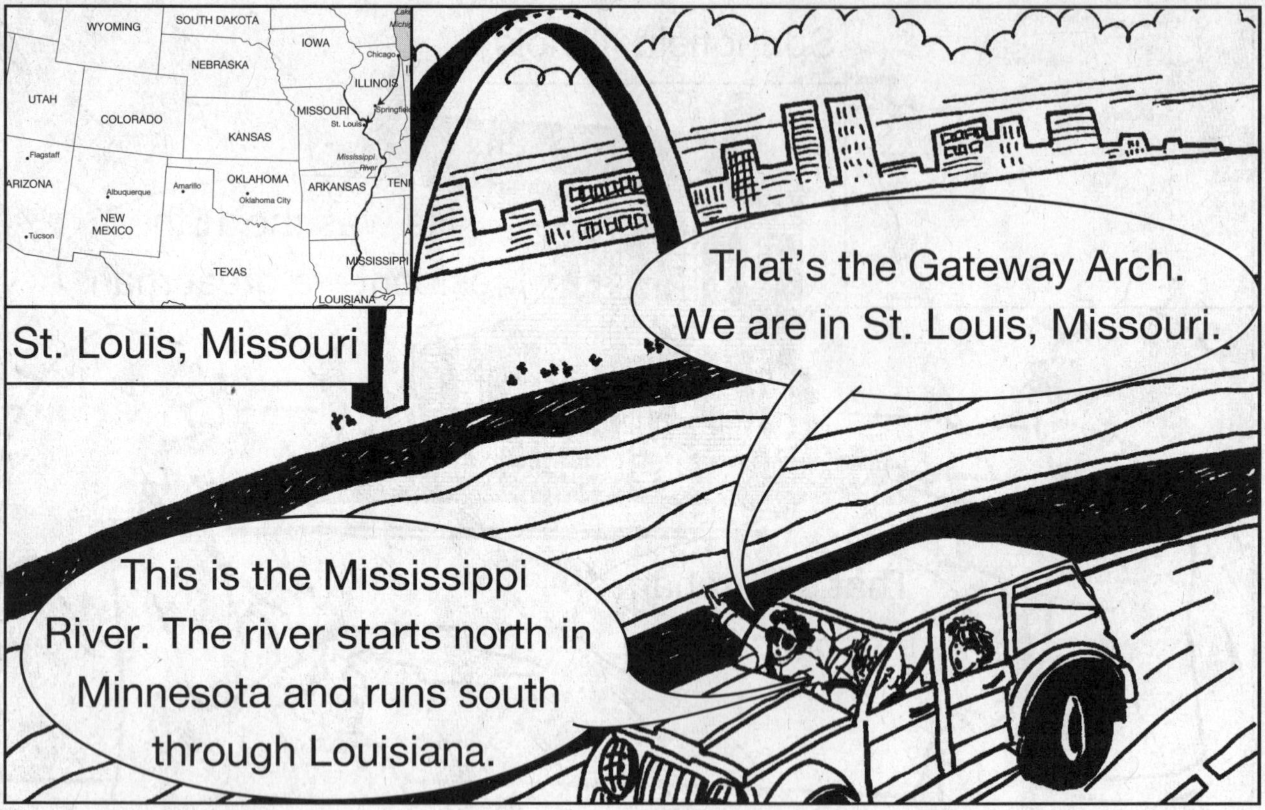
WYOMING
SOUTH DAKOTA
IOWA
NEBRASKA
ILLINOIS
Chicago
UTAH
COLORADO
MISSOURI
St. Louis
KANSAS
Flagstaff
ARIZONA
Albuquerque
Amarillo
OKLAHOMA
Oklahoma City
ARKANSAS
NEW MEXICO
Tucson
TEXAS
MISSISSIPPI
LOUISIANA
St. Louis, Missouri
That's the Gateway Arch. We are in St. Louis, Missouri.
This is the Mississippi River. The river starts north in Minnesota and runs south through Louisiana.

The arch is so big! Are we going to the top?
Yes!
Oh, boy!

What a view!
It's cloudy again. But not as cloudy as in Chicago.
It's warmer and sunnier here.

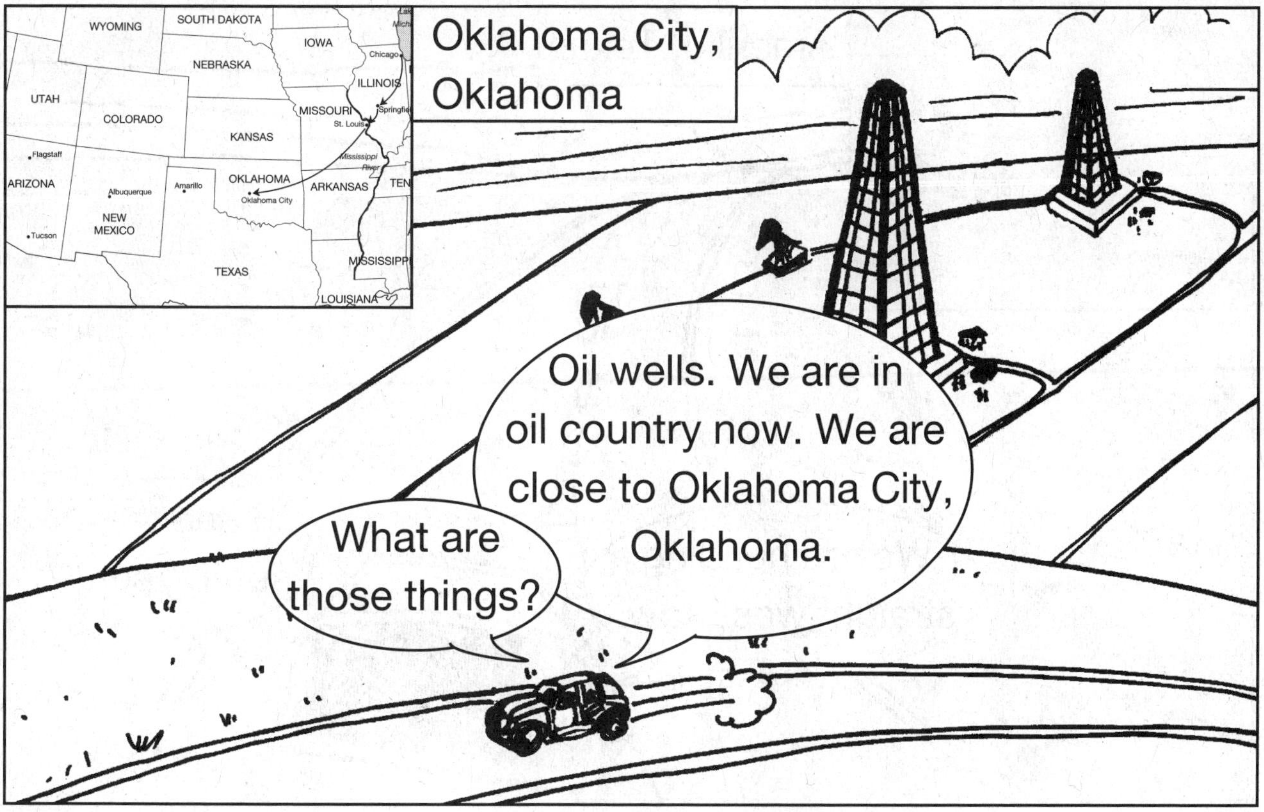
WYOMING
SOUTH DAKOTA
IOWA
NEBRASKA
Chicago
ILLINOIS
UTAH
COLORADO
MISSOURI
Springfield
St. Louis
KANSAS
Mississippi River
Flagstaff
ARIZONA
Albuquerque
Amarillo
OKLAHOMA
Oklahoma City
ARKANSAS
NEW MEXICO
Tucson
TEXAS
MISSISSIPPI
LOUISIANA
Oklahoma City, Oklahoma
Oil wells. We are in oil country now. We are close to Oklahoma City, Oklahoma.
What are those things?

I see
horses, too!
It's a lot sunnier
and drier.

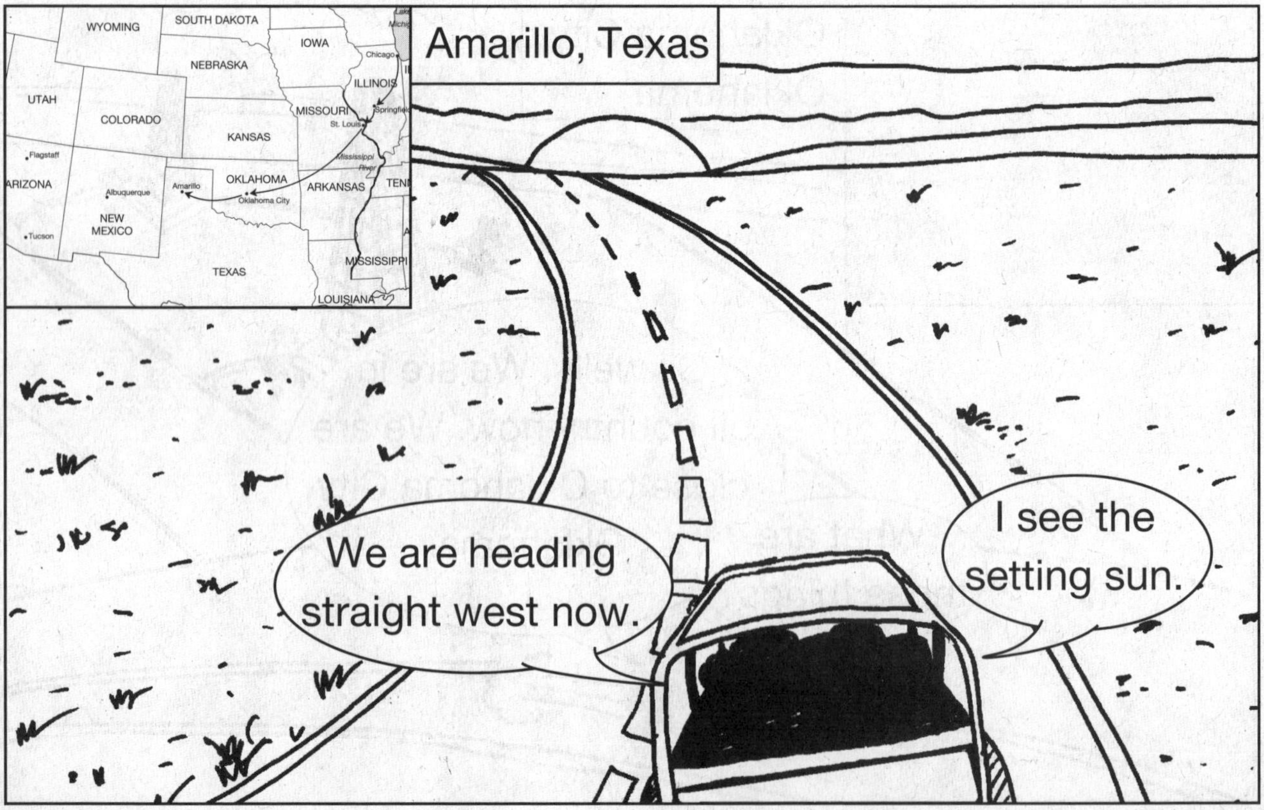
Amarillo, Texas
WYOMING
SOUTH DAKOTA
IOWA
NEBRASKA
Chicago
ILLINOIS
UTAH
COLORADO
MISSOURI
Springfield
St. Louis
KANSAS
Flagstaff
Mississippi River
ARIZONA
OKLAHOMA
ARKANSAS
Albuquerque
Amarillo
Oklahoma City
NEW MEXICO
Tucson
TEXAS
MISSISSIPPI
LOUISIANA
We are heading
straight west now.
I see the
setting sun.

US
40
See the highway sign? We are on U.S. 40.
Hey! A rabbit!
YIKES!

Look at the cows! *Moooooo!*
Finally, we're seeing Texas Longhorn cattle.

We are near Amarillo, Texas.
It's very pleasant and sunny.

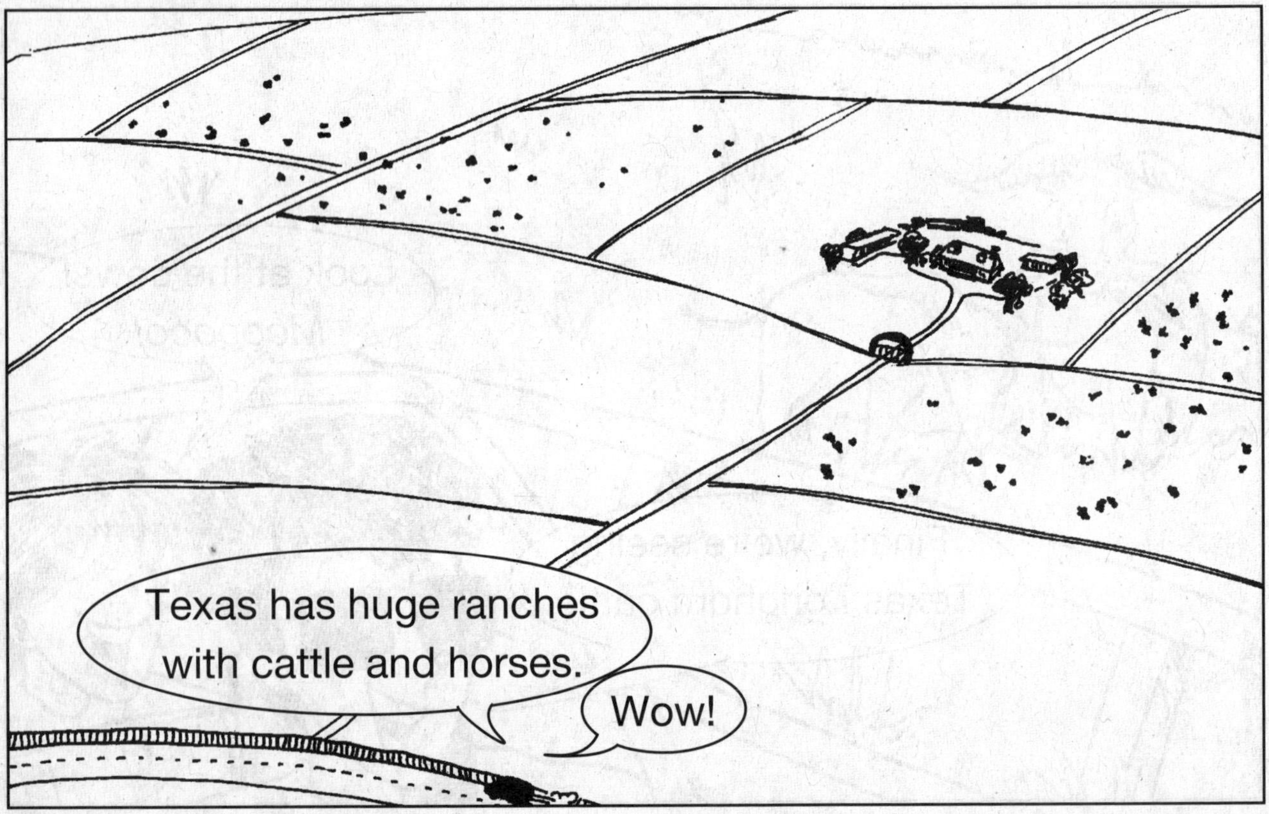
Texas has huge ranches with cattle and horses.
Wow!

Albuquerque, New Mexico
WYOMING
SOUTH DAKOTA
IOWA
NEBRASKA
Chicago
ILLINOIS
UTAH
COLORADO
MISSOURI
Springfield
St. Louis
KANSAS
Flagstaff
Mississippi River
ARIZONA
Albuquerque
Amarillo
OKLAHOMA
Oklahoma City
ARKANSAS
NEW MEXICO
Tucson
TEXAS
MISSISSIPPI
LOUISIANA
We are near Albuquerque, New Mexico.
Whooooo Whooooo
Look, a train.

What are those drawings on the rocks?
Wow!

Look, but don't touch. They are very old.
These are wall paintings of an animal hunt done by Native Americans. Native Americans have lived in Arizona for hundreds of years.

WYOMING
SOUTH DAKOTA
IOWA
NEBRASKA
Chicago
ILLINOIS
UTAH
COLORADO
MISSOURI
Springfield
St. Louis
KANSAS
Flagstaff
Mississippi River
OKLAHOMA
ARKANSAS
Albuquerque
Amarillo
Oklahoma City
NEW MEXICO
Tucson
TEXAS
MISSISSIPPI
LOUISIANA
Flagstaff, Arizona
Mountains! We are near Flagstaff, Arizona.
There's snow, but it's warm. I like that.
We're going to go south on Highway 10. Keep an eye out for it.

What is that?
Casa Grande. In Spanish, it means *big house.* Native Americans lived here a long time ago.

It's warm now and sunny almost all the time.
We're only three days out of Chicago! It's hard to believe it's still January.

It's Sunny in Arizona

A World of Cubic Animals

What did you do in school today, Manuel?
We made shapes out of cubes.
Neat!
Shhh! Don't tell Mom, Ruffy.
Munch Munch

See, the two different shapes have the same volume, six cubes.
I see. You count the cubes to find the volume.

All those different shapes…
All those different cubes…
ZZZZZZZZZZZZZZZZZZZ

In the dream…
What happened to you, Ruffy? You're made of cubes!
RUFF RUFF

There's Farmer Jones...
Hi, Manuel.
Hi!

Look at Bell!

She's made of cubes, too!
Hi, Manuel!

Neat! I know everyone's volume now because I can count the cubes!

I wonder if the animals
in the zoo are that way, too.
Let's go see.
ZOO

There's
the wolf.

And there's the snake.
It's really long. I wonder what its volume is. What will the giraffe be like? Let's see...

It's made of cubes, too!
Can you count them?

Look at the baby elephant. It has a large volume.

Ruffy, I wonder what I would look like as a cubic person.

I AM A CUBIC PERSON!
!

What's my volume, Ruffy? Ruffy? Ruffy! Come back! It's me!
Help!

Okay! Enough!
I want to be a
person again!

It's not working.
I can't get back!

Wake up, dear. Wake up.
I can't get back! I can't get back!
?

Mom, is that you?
You were having a dream.

What? Cubes? No, silly, you must have dreamed that. Go back to bed.
Am I made of cubes?

I'm glad we're not made of cubes, Ruffy. But it does make finding volume easy!

The Martians

On the planet Mars...

"I need volunteers for a mission to Earth," announced the famous Martian scientist named Brainy. "We must study Earthlings! Who wants to go?"

"We all do!" shouted the Martians.

"There's Earth now," said Brainy.

"We'll be landing in the United States," said the Captain.

"Dopp and Zorka, it's time you transformed into Earthlings," reminded Brainy.

As soon as the Martians landed, they saw an interesting looking building. Many small Earthlings were in and around it. Some were sitting at desks, reading and talking together. Others were playing outside, running and jumping. They were so busy!

Dopp and Zorka looked at the Earthlings and wondered where they got all that energy! They wondered what the Earthlings ate to keep their bodies going. They agreed that this would be a good thing to study.

A day later, Dopp and Zorka returned to the spaceship.

"They're back," called Captain.

"Did you collect your data and find out what you needed to know?" asked Brainy.

"Yes," they replied.

"And, did you write a report?" asked Brainy.

"Yes, here it is," said Dopp and Zorka.

"We found out that some Earthlings have lots of energy and others do not," said Dopp.

"Earthlings eat a lot of different foods," said Zorka.

"It seems that the Earthlings who ate certain kinds of food were healthy and had lots of energy," said Dopp.

"Yeah, but some Earthlings ate other kinds of food and they were tired and seemed unhealthy," said Zorka.

"Why would they eat food that would make them tired and unhealthy?" asked Brainy.

"We didn't find out," said Dopp.

"But next time we go to Earth, we're going to find out!" said Zorka.

Tensland

Once upon a time, there lived a woman who had two daughters. Their names were Rosie and Hildie.

Both girls were hardworking and clever.

If Rosie found a bird with a hurt wing, she cared for it until it could fly again. If she saw a plant that was thirsty, she watered it.

Hildie enjoyed reading quietly under her favorite tree. She liked to sit and think about important things like clouds and rainbows.

One cold winter morning, their mother said, "Rosie and Hildie, please go to the well and fetch water for the day."

It was a very, very cold day! When the girls reached the well, they found a layer of ice on top of the well water. Rosie leaned over to break the ice and started to fall. Hildie grabbed hold of Rosie's foot, but both girls tumbled down the well.

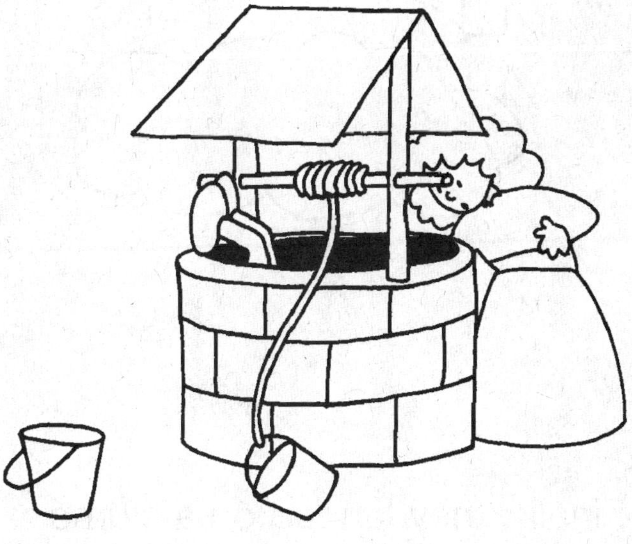

They fell down,

down,

down.

Finally, they landed on a clump of soft green grass.

When they looked around, they discovered a beautiful land called Tensland. It was filled with strange plants and animals.

Rosie and Hildie began walking through Tensland. Soon, they came upon a crowd of cubes.

"Help us!" cried the cubes. "We don't know how many we are because we are all jumbled up. Please put us in order."

"We will be glad to help you," said Rosie and Hildie. "First, we will count you one at a time. Then, we will put you in a neater pile."

The two sisters started counting together. After a while, Rosie said to Hildie, "So far, we have counted 24 cubes."

"No," Hildie replied. "We have counted 25 cubes."

"Are you sure?" Rosie asked. "Did you skip a number? Maybe I counted the same cube twice. Oh, no! I'm getting so mixed up. Let's start over."

Hildie said, "What if we make a mistake again? Also, counting one by one takes too long. There must be a better way to help the cubes."

The girls thought and thought. "I have an idea," said Rosie. "Let's group the cubes by tens. Then, they will be organized and easier to count."

So, the girls put the cubes in groups of ten. When they were finished, a couple of cubes were left over.

The cubes were very pleased. "Thank you so much," said the cubes. "Now we know how many we are altogether. We also feel much better because you stacked us in groups."

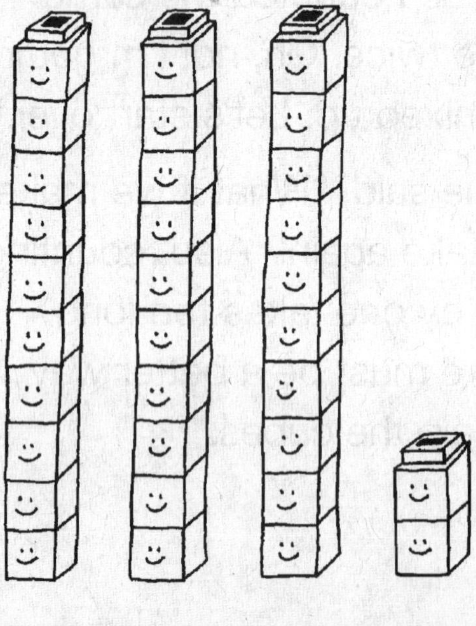

Next, the girls came upon a pile of apples.

"Help us, please!" cried the apples. "We are going to market, but no one will buy us this way! Please put us in bags!"

So, Hildie and Rosie put the apples in bags by tens.

"There you are," said Hildie. "All the bags have ten apples each. These four apples are left over."

"Thank you, kind girls!" cried the apples. "Now, someone will buy us at the market."

Next, the girls came to a huge pile of sticks. "Please help us!" cried the sticks. "We are a mess! Please get us organized!"

"No problem," said Rosie and Hildie. "We will group you first by tens and then by hundreds." So, the girls worked together to put rubber bands around every ten sticks.

Then, Rosie and Hildie tied ten groups of ten together to make bunches of hundreds.

When they finally finished, the girls had two big bunches of 100 sticks each, six bundles of ten sticks each, and five sticks left over. Once the girls had grouped the sticks, they were easy to count. Rosie and Hildie told the sticks how many they were altogether.

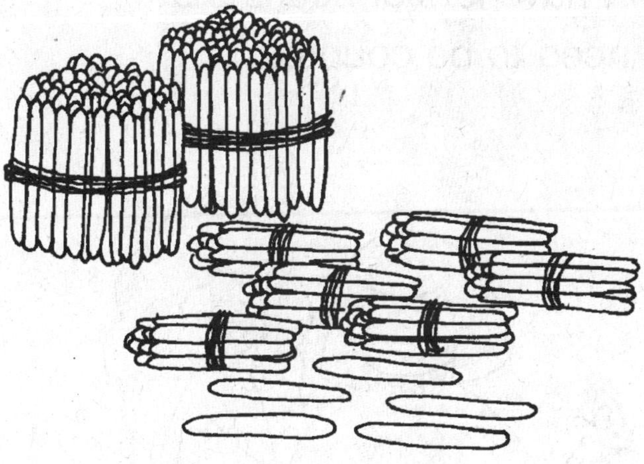

"Thank you, kind girls," said the sticks. "Now, we know how many we are!"

Rosie and Hildie were eager to meet some people in Tensland. They kept walking until they came to a small hut. Rosie and Hildie knocked on the door.

A very old woman came to the door. The girls were hungry and asked the woman for food.

"I will give you food if you will work," said the old woman. "I have lots of beans that need to be counted."

"We are good at counting and will be happy to help you," said the girls.

So, the old woman gave the girls a bowl of beans to count.

Rosie and Hildie got some small cups and put ten beans in each cup. Then, they stacked ten cups to make 100 beans. The girls worked for a long time. They checked their counting carefully. They didn't want to tell the old woman the wrong number.

At last, Rosie and Hildie had five stacks of ten cups. Each of these cups was filled with ten beans. They also had one stack of three cups, each filled with ten beans. There were seven loose beans left over. Now, they could count the beans easily. They told the old woman how many beans there were.

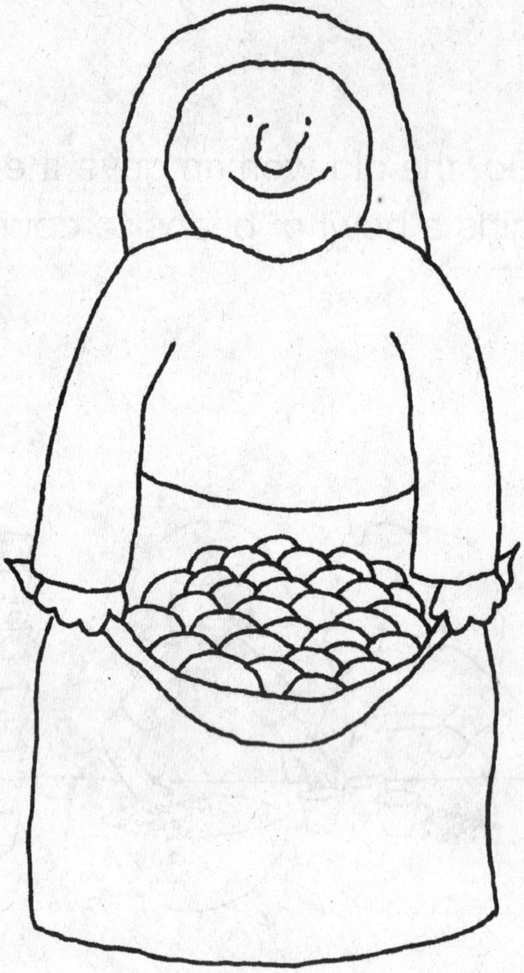

The old woman was very happy with the girls' counting. Every day, she gave them more things to count. Every day, they grouped things by hundreds, tens, and ones. Rosie and Hildie always counted correctly.

The old woman was very kind to Rosie and Hildie. But, they missed their mother very much. And, of course, their mother missed her daughters, too.

The old woman told the girls the way to return home. Before they left, she rewarded them for the many good deeds they had done in Tensland.

"Every morning when you awake, you will find a gold coin under each of your pillows," said the old woman.

Rosie and Hildie said, "You do not owe us such a wonderful reward." The old woman told the girls how much they deserved it. They thanked her and began their trip home.

Rosie and Hildie's mother was very happy to see them. She was excited to learn about the gold coins. But she wanted the girls to keep only enough gold for the small family to live well.

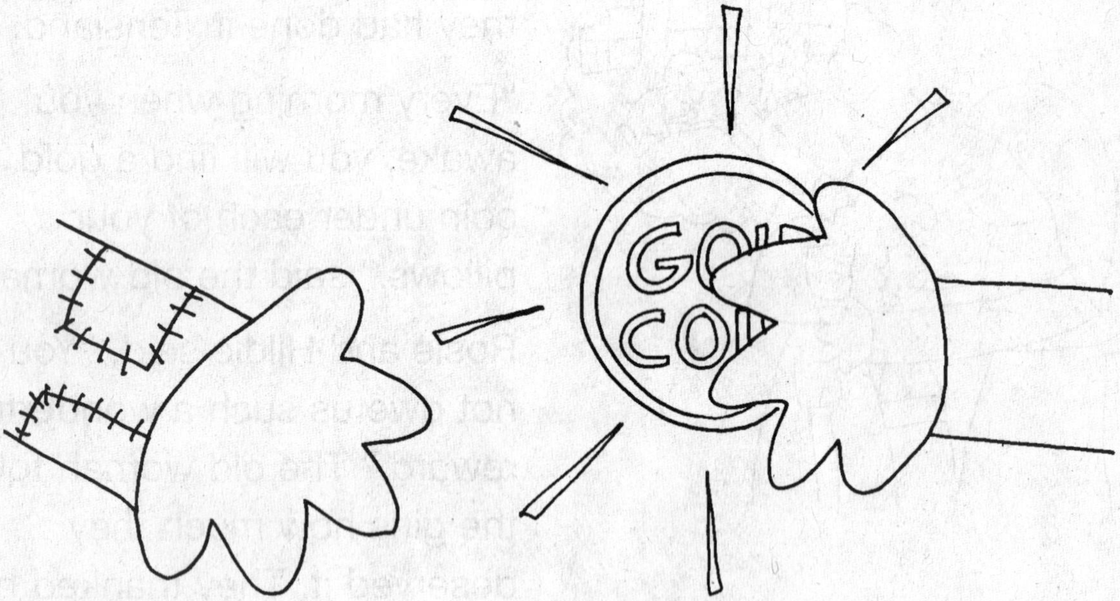

Rosie and Hildie gave their extra gold coins to needy families in the village. And Rosie, Hildie, and their mother all lived happily ever after.

Buried Treasure

Stay in line, class.
Boy, I hate museums. They're boring!
Not me. I love them.
MUSEUM OF NATURAL HISTORY
XZ12

What's this, Helen?
I don't know, Johnny.
ARTIFACTS

Nobody knows, Johnny. It was discovered about 20 years ago at an ancient Navajo site that is not too far from here.

It reminds me of the Mr. Origin map we made in class.
Yes! It has a line with marks and different things are along the line.

See, the marks are equally spaced.
I think these circles represent things!

But I wonder *what* things?
Maybe the things are buried treasure!

Where is the buried treasure? Where do we start? How far apart are the marks?
I don't know yet. Let's copy the map, and maybe we will figure these markings out later.

Later, on the bus...
What do you have?
We think it's a map.

What does the map show?
Don't be silly!
Buried treasure!

The scientists don't know what that map is. How could you?
We will find out!
Ha! Ha! Ha! Ha! Ha! Ha! Ha!
Everyone sit down and buckle up. We're going to be leaving.

Later on, after school...
How are we going to find the buried treasure? We don't know where to start. Maybe we can use Mr. Origin like we did in class.
Look, there are three triangles that look like mountains. The sun is behind the middle mountain.

Maybe the rectangle is where we put Mr. Origin.
But what is that rectangle, and where is it? I give up! Let's go to my house to watch TV.

On the way to Helen's house...
I wonder what that map means...
The sun is setting behind the mountains.

Look, the map in the museum has the sun in the same spot.
Maybe that's the clue to finding that rectangle.

Late the next day, Helen and Johnny bike to the foot of the mountains. They climb up a bit.
Hiking is hard work!
Look, there's a bunch of big rocks. They are shaped like rectangles. They look like the rectangle on our map.

Look, the shadow of the mountain is pointing to that rock.
That rock could be the rectangle on the map. And look, next to the rock is an old Navajo trail. That could be the line on our map.

Maybe Mr. Origin can help us now. We will put Mr. Origin on the rectangle.
But how far apart are the marks on the map?

At Johnny's house that night...
I have an idea about those marks on the map.
What, Mr. Eagle?

Those marks could be a day's walk apart.
People can walk about 2 miles in an hour and walk for 8 hours a day.
I think that would have been a natural way to measure distance at that time.
I get it.

That would be 2 miles 8 times. So, our daily distance would be 2 miles + 2 miles + 2 miles + 2 miles + 2 miles + 2 miles + 2 miles + 2 miles.
That's 16 miles in a day!

So, the first circle to the right of the rock is one day away. The distance is 16 miles to the right.

We can't walk that far!
No. But maybe we can bike it. Maybe we could get my mother to go with us!

Early the next morning...
Here's the rock, and there's the trail.
The rock can stand for zero. I'll keep track of the distance. Let's go.

About 16 miles later...
I don't see anything.
Neither do I. Let's walk our bikes and look around.

Look at those rocks!
They almost make a circle. That might be the circle along the line on our map.

I bet this is the spot for our buried treasure! Let's dig!
No. We should call the scientists at the museum.
Good idea!

The next day...
Fantastic, kids! We never would have figured this out. Look at all this great stuff! This is all history! What a find! There are pottery, jewelry, and so many other rare objects. These add so much to our collection.

And there are more circles on the map! You kids should get a reward. We are so proud of you that we're going to name the collection in your honor.

The Helen Rizza and Johnny Eagle Collection of Navajo Art
We couldn't have done it without Mr. Origin.
We are a great team, Johnny—you, me, and Mr. Origin.